Einstein's

Spooky Action at a Distance

Introduction

Albert Einstein was a German-born theoretical physicist who developed the theory of relativity, one of the two pillars of modern physics (alongside quantum mechanics). He is known to the general public for his mass energy equivalence formula $E = mc^2$, which has been dubbed "the world's most famous equation". He received the 1921 Nobel Prize in Physics "for his services to theoretical physics, and especially for his discovery of the law of the photoelectric effect", a pivotal step in the development of quantum theory. He subsequently realized that the principle of relativity could be extended to gravitational fields, and published a paper on general relativity in 1916 introducing his theory of gravitation.

He continued to deal with problems of statistical mechanics and quantum theory, which led to his explanations of particle theory and the motion of molecules. He also investigated the thermal properties of light and the quantum theory of radiation, the basis of laser, which laid the foundation of the photon theory of light. In 1917, he applied the general theory of relativity to model the structure of the universe.

Einstein moved to Switzerland in 1895 and renounced his German citizenship in 1896. After being stateless for more than five years, he acquired Swiss citizenship in 1901, which he kept for the rest of his life. Except for one year in Prague, he lived in Switzerland between 1895 and 1914.

In 1933, while Einstein was visiting the United States, Adolf Hitler came to power. Because of his Jewish background, Einstein did not return to Germany. He settled in the United States and became an American citizen in 1940. On the eve of World War II, he endorsed a letter to President Franklin D. Roosevelt alerting FDR to the potential development of "extremely powerful bombs of a new type" and recommending that the US begin similar research. This eventually led to the Manhattan Project.

Long before the Manhattan Project, in a 1905 paper, Einstein postulated that light itself consists of localized particles (quanta).

Einstein's light quanta were nearly universally rejected by all physicists, including Max Planck and Niels Bohr. This idea only became universally accepted in 1919, with Robert Millikan's detailed experiments on the photoelectric effect, and with the measurement of Compton scattering.

Quantum theory emerged in part in heated clashes between Niels Bohr and Albert Einstein. The theory posed a challenge to the very nature of science, and it severely straining the relationship between theory and the nature of reality.

Niels Henrik David Bohr was a Danish physicist who made great foundational contributions to understanding atomic structure and quantum theory, for which he received the Nobel Prize in Physics in 1922. Bohr was also a philosopher and a promoter of scientific research. Bohr was born in Copenhagen, Denmark, on 7 October 1885, the second of three children of Christian Bohr, professor of physiology at the University of Copenhagen, and Ellen Adler Bohr, who came from a wealthy Danish family prominent in banking and parliamentary circles. He had an elder sister, Jenny, and a younger brother Herald. Jenny became a teacher, while Herald became a mathematician and footballer who played for the Danish national team at the 1908 Summer Olympics in London.

Niels was a passionate footballer as well, and the two brothers played several matches for the Copenhagen-based Akademisk Boldklub (Academic Football Club), with Niels as goalkeeper.

Spooky Action at a Distance

For a moment, imagine 2 particles that are like spinning coins. Now imagine that these two coins are two electrons atoms that were created from the same event and then quickly moved apart far away from one another.

Quantum mechanics theory says that since these 2 electrons were created at the same time, and from the same event, they are entangled forever, and most of their properties are forever linked wherever they are.

For example, if we suddenly stop the first spinning coin and if it shows heads, the second coin will instantaneously become tails, since the two coins are linked and entangled forever.

However, Einstein said that somehow the destiny of the two coins, whether or not they ended up heads or tails, was already fixed in time long before we observed them. He said although it seemed that the coin was deciding to be, say heads, at the moment of observation, but actually that decision was taken long before that. It was just hidden from us.

Einstein said that such quantum atoms are nothing like spinning coins. They are more like a pair of gloves, a left and a right hand, and they are separated into boxes or drawers. And we do not know anything, like which drawer contains which glove until we actually open one of the drawers, but if we open the drawer and find a right-handed glove, then immediately we will know that the other drawer has the left-handed glove.

So fundamentally, this requires no "spooky action at a distance," at all, because neither glove has changed by the act of observation. Both gloves were either right or left-handed from the start, from the beginning, and the only important detail that changed was simply our knowledge of them.

So, which is the true description on reality? Some may think that Bohr's coins, which become real only when we observe them, and then they magically communicate with one another? Others prefer Einstein's gloves, which are hidden in a box, but are definitely left or right-handed from the start. So is there a true reality, as Einstein said, or not, as Bohr maintained?

Bohr said that the atom particles seem to communicate with each other faster than the speed of light. But Einstein said that it was impossible. However, in recent years many experiments have shown that subatomic particles really are entangled. They can subtly and instantaneously influence each other across space and time.

Bird's Navigation and Quantum Theory and Entanglement

For decades, exactly how many birds navigated with such accuracy was one of the greatest mysteries in biology. So the most recent discovery has caused a sensation. In the past few years, one species of birds has helped create a scientific revolution.

I was one of many scientists who was shocked to discover that it navigates using one of the strangest tricks in the whole of science. It utilizes a quirk of quantum theory mechanics, one that puzzled even the greatest of physicists from Richard Feyman to Albert Einstein himself.

So you might be surprised to discover the identity of this mysterious creature. Say hello to the Quantum Robin. The European Robin. Every year she migrates from northern Europe to the tip of Spain and back. Biologists were trying to solve the mystery of how she does it.

They found themselves in my world, the strange world of quantum mechanics. Quantum mechanics describes the very weird behavior of subatomic particles. Down in this realm of the very small, we have to abandon common sense and intuition.

Instead, this is a world where objects can spread out like waves. Quantum particles can be in many places at once and send each other mysterious communications.

So biologist set out to find out how the robin finds its way, and their data more and more pointed towards Quantum particles as the only explanation that could bring all the different results together.

Biologist discovered that robins navigate by the earth magnetic field. When a photon enters the robin's eye, it creates what is called an entangled pair of electrons.

Here is how it works. Each electron has two possible states. For simplicity I am choosing to call them green and red. Now here is the weird thing, until I measure it, it is neither red nor green, but both at the same time. Think of the electrons like a spinning discs with red and green. They are simultaneously red and green.

But when I fire a dart at the spinning disc, I can force the first electron to be one color or the other. So far it is just a game of chance. I do not know what I will get until I try it. So if the dart lands on a red triangle, then I know my first electron is red. Suppose I now measure the second electron. You would think I have a 50-50 chance of hitting (getting) red or green.

After all that is what you would expect in the normal everyday world, but you would be wrong. In quantum theory entanglement, the electrons are mysteriously linked. For example, if I get red on the first, I will always get red on the second. It is not a game of chance anymore. It is as if the first electron is telling the second what to do.

That is why Einstein called it spooky. The electrons seem to know that they should both have the same color, no matter how far apart they are. The really important part, is that the two electrons need not be the same color. They can be entangled in a different way, so if the first electron is red, the second on is always green.

It seems that this odd and mysterious connection is the ultimate secret of the quantum robin's compass, because the direction of the earth's magnetic field can influence the outcome. Near the equator they may be more likely to be red and red.

But near the earth's pole they may be more likely to be red and green. And that is the vital factor that finally tips the balance of the robin's chemical compass within the robin's eyes.

Tiny variations in the earth's magnetic field change the way electrons in the robin's eye are entangled, and that is just enough to trigger her compass.

Now finally we can see how something as weak as the earth's magnetic field can tip that balance one way or the other. If the message changes the chemical reaction tips a different way changing the robin's compass reading. If it looks like it is a fundamentally quantum mechanical phenomenon in birds, it would be one of the first in biology.

Biologists better get used to the weirdness of physics. The robin is truly navigating by the spooky quantum entanglement. To see subtle quantum effect even in a controlled sterile environment of physics lab is really difficult.

And yet here is the robin doing it with ease. These experiments are real and verifiable and yet even though I am seeing them with my own eyes, I still find it hard to believe.

By the late 1930s (with World War II looming), there was very little interest in Einstein's and Bohr arguments and questions. Therefore, the battle to unlock the mystery of the entangled universe, and to understand the nature of reality was deadlocked. Before the start of the World War, most of the scientists fled to America, after that, Quantum Mechanics led to a deep and profound understanding of electronics and semiconductors that helped create the modern electronic age.

The scientists invented lasers, and revolutionized communications, & built breath-taking new medical devices, and made breakthroughs in nuclear power and science.

Quantum mechanics was so successful that most scientists ignored Einstein's objections. It simply did not matter to them at all, because it worked and money was flowing in like rain. They even coined a phrase for it, "Shut up and calculate."

The price for this technological success was that Einstein's and Bohr's debate on the reality and truth of the quantum world was simply pushed and brushed under the carpet. And amidst all this success and pragmatism, there were a few scientists that still worried what it all meant.

However, as the 1950s rolled into the 1960s, one lone physicist found a way to settle the argument once and for all. His name was John Bell (1928-1990). Bohr was inconsistent, unclear, and wilfully very obscure but right. And Einstein was consistent, clear down-to-earth, kind, and wrong.

John Bell was not well-known at all to the general public, but to scientists like me, he is, well, a hero. He was a truly an original thinker with real courage in his convictions, and the story of his rise to become one of the greats is made even much more remarkable when you consider how his life started.

John Bell was born in Belfast, in the 1920s. Belfast is the capital and largest city of Northern Ireland, standing on the banks of the River Lagan on the east coast of Northern Ireland. It is the 12th-largest city in the United Kingdom. John Bell's family was a poor working-class family. His father was a kind man and a horse dealer. Everyone loved him. Bell's father struggled to get him into Queens University, Belfast, to study physics.

John Bell was the only one in his family to even finish school. This made John insatiably curious, fiery, and stubborn. For many years he worked at Britain's Atomic energy Research Center at Harwell, who built this experimental nuclear reactor called DIDO, which was a materials testing nuclear reactor.

It was used enriched uranium metal fuel, and heavy water as both neutron moderator and primary coolant. There was also a graphite neutron reflector surrounding the core.

At work, John started pondering the deep and crazy questions that quantum mechanics raised. Does the quantum world only exist only when it was observed? Or was there a deeper truth out there, waiting to be discovered?

John began to wonder if there was a problem at the heart of quantum mechanics. So he decided to try and to resolve the questions at the heart of quantum physics.

But how does one check if something is real? If something is or is not there, all without looking?

How can one look behind the curtain without pulling it open? John Bell came up with a brilliant way of doing exactly that. I think this is one of the most clever and ingenious ideas in the whole of physics. It is certainly one of the most difficult to understand and to explain, but I am going to try and have a go at it. I am going to use an analogy.

I am going to play a game of cards, but it is one for the highest possible stakes, the nature of reality itself. The card game is against the mysterious evil quantum dealer.

The cards he deals me represent any subatomic particles, or even quanta of light, photons. The game we will play will ultimately tell us whether Albert Einstein or Bohr was right.

Now the rules of the card game are deceptively simple. The dealer is going to deal me two cards, face down. If they are the same color, I win, if they are different colors, I lose.

In the first hand, I have a red, so I need another red to win. That is black, I lose. I tried again, and again, a 1000 times, but I kept losing. 1000 in row and the same opposite color cards.

I think I know what the dealer is doing here, clearly the deck has been rigged in advance so that every pair come out as opposite colors. But there is a simple way to catch the dealer out. So what can I do now? I can change the rules of the game.

This time if they are the opposite color, I win. But once again, every time my evil quantum opponent won. But again I can see what the crafty dealer could have done. Maybe while I was not looking he has rigged the cards again, so it always lands in his favor. Remember what Einstein thought was really happening in particles entanglement.

He said that just like the gloves that were already placed in the box, so the evil dealer stacked the cards before we played. But Niels Bohr's idea was very different. He said red and black do not even exist until you turn them over.

Bell's was a master of experimental physics. He came up with a way of deciding once and for all who was right, Einstein or Bohr. This is how he did it.

I will explain his probability equation as such. I am now not going to tell the dealer which game I want to play, same color wins, or different color wins, until after he is dealt the cards.

Now, because he can never predict which rules I am going to play by, he can never stack the deck correctly. Now, he cannot win. Or can he?

So now different colors win. But again, the dealer won. Ok, now if they are the same color I win. But the dealer won again.

This gets to the very heart of Bell's idea. If we now start playing and I win as many as I lose, then Einstein was right, the dealer is just a trickster. Reality maybe tricky, but it does have an objective existence. But what if I lose? Well then, I am forced to admit that there is no sensible explanation.

Each card must be sending secret signals to the other card, across space and time, in defiance of logic. I am forced to accept that the fundamental, quantum level, reality is truly a mystery. Bell reduced this idea into a single mathematical equation that tell us once and for all what seemed unanswerable. How reality really is.

$$P(a,c) - P(b,a) - P(b,c) <= 1$$

John Bell published his idea in 1964, and the extraordinary thing is, at that time, the entire physics community ignored him. There was total silence. It seems the world simply was not ready or they just chose to ignore him.

Perhaps it was because his equation seemed untestable, or nobody thought it was worth investigating. But that was about to change, and the change would come from a much unexpected place.

America was in crisis, over Vietnam, Watergate, Feminism, the Black Panthers, and while all this was going on, a small group of hippy physicists were working at the University of Berkeley in California.

They did all the hippy things. They smoked dope, they popped LSD, and they debated things like Buddhism and telepathy. And they loved quantum mechanics.

In its weird version of reality they saw parallels with their own esoteric beliefs. Their hippy, new-age style physics also caught the attention of the public who read their crazy, hippy books. That mixed quantum mechanics with Eastern mysticism. They wrote books like the Tao of Physics, The Dancing Wu Li Masters, any my personal favorite, Space-time and Beyond: Towards an Explanation of the Unexplainable.

But more importantly for our story, the story of quantum mechanics, these hippy physicists also turned their attention to Einstein's now famous thought experiment (the gloves) and what it told us about the nature of reality.

They saw Niels Bohr's secret signaling as proof that physics supported their own ideas, because if two particles could spookily communicate across space and time, then ESP, telepathy, and clairvoyance were probably true as well.

If only they could prove it really existed. Then, in 1972, they realized that, with a bit of mathematical skill, they could take Bell's equation and experimentally test it.

One of their group, John Clauser, borrowed some equipment from the lab he was working in and set up the first genuine and ultimate test of quantum mechanics.

Over the next few years, it was improved by a team led by Alain Aspect in Paris, making its results more reliable.

Over ten years after Bell first proposed his equation, finally it could be put to the test. This is a modern version of the experiment first carried out by John Clasuer and then Alain Aspect.

In their experiment, a crystal converts laser light into pairs of entangled light quanta, photons, making two very precise beams. These photons are passed around and bent back again, until they pass through these detectors.

The two photons are like the two cards the evil dealer places in front of me. We will measure a property of the photons called polarization. Which is equivalent to the color of the playing cards in my game. So, for instance, winning with two matching red cards might be the same as two photons with matching polarization. But since this is quantum mechanics it is more complicated than my simple card game.

Their experiment also allowed them to measure a second property of the photons as well. It is equivalent to me not only trying to guess the color of the face of the cards, but also trying to guess the color of the back of the cards.

Okay, so we now are going to switch on the laser and start the experiment. So first number (Total games played: 5184) and (Expectation value: 0.55903), and (Uncertainty: .0097529). This gives me the number of photon pairs coming through the experiment. That is equivalent to the pairs of cards in my game.

When the numbers were graphed, if the graph was dropping down, that showed me the probability to win, and that I am guessing right. The more photons, the more accurate it becomes. I will stop at an uncertainty of about 1%. So the final answer is .55903.

If I put that into Bell’s equation

$$P(a,c) - P(b,a) - P(b,c) <= 1$$

I now need to run the experiment three more times. Each run is now like a different set of rules for the quantum dealer. And when I add them up and get the answer (using the above equation). If it is <= 2 then Einstein was right. If it is greater than two, then Bohr was right.

Okay, so now for the second setting. Just remember what the experiment will show. If the numbers come out less than two, then it is proof the dealer has been stacking the deck. This was Einstein's view.

Okay, so the number I get this time is .82. Now, reset for run three. Change the dial again. But if the result is greater than two, then the deck cannot be stacked and something else is at work. For the run three result is -.59 (expectation value)

$$.55 + .82 + .59$$

And finally, run four (change the dial again). This last number will finally reveal the world follows common sense or something much more bizarre. Okay, so our final result is in, and it is .56. So if we turn the laser off.

$.55 + .82 + .59 + .56 = 2.53$ which is > 2

Absolute proof that Albert Einstein was wrong and Niels Bohr was right. The significance of this result is simply enormous. Just remember what it means, Einstein's version of reality cannot be true. No amount of clever jiggery-pikery with our experiment can cheat nature. The two entangled photons' properties could not have been set from the beginning, but are summoned into existence only when we measure them.

Something strange is linking them across space and time. Something we cannot explain or even imagine other than by using mathematics. And weirder, photons do only become real when we observe them.

In some strange sense it really does suggest the moon does not exist when we are not looking. It truly defies common sense. No wonder towards the end of his life, Einstein wrote, "All these 50 years of conscious brooding have brought me no nearer to the question - what are light quanta? Every Tom, Dick, and Harry thinks he knows it. But he is mistaken."

The experiment only confirms this, whatever is happening, we just do not understand it. But it does not mean we should stop looking. While it is true that Einstein's dream of finding a reasonable, common sense explanation was shattered for good, my own personal view is that this does not necessarily banish physical reality. Like

Einstein, I still believe there might be a more palatable explanation underlying the weird results of quantum mechanics. But one thing is clear - Whether there are physical, spooky connections, whether there are Parallel Universes, whether we bring reality into existence by looking, whatever the truth is, the weirdness of the Quantum world, will not go away.

However, Einstein was rarely wrong when it came to science, but this time he was wonderfully far off the mark. Numerous experiments have shown that the effect he contemptuously dismissed as "spooky action at a distance" is a fundamental aspect of nature.

However, I still believe in Einstein. How can something communicate faster than light?

Faster-than-light communication and travel are the conjectural propagation of information or matter faster than the speed of light. The special theory of relativity implies that only particles with zero rest mass may travel at the speed of light.

In special relativity, it is truly impossible to accelerate an object to the speed of light, or for a massive object to move at the speed of light. However, it might be possible for an object to exist which always can move faster than light.

The hypothetical elementary particles with this property are called tachyonic particles. Many attempts to quantize them failed to produce faster-than-light particles. Moreover it illustrated that their presence leads to an instability.

Einstein's Quantum Riddle

Is reality an illusion?

Could something here mysteriously affect something on the other side of the universe?

A century of discoveries in physics reveals a strange, counterintuitive micro-world of atoms and tiny particles that challenges our intuitive understanding of the world we see around us. It is known as Einstein's Quantum Mechanics.

This strange Einstein's theory has enabled us to develop the remarkable technologies of our digital age.

But it makes very troubling prediction, it is called quantum entanglement.

Entanglement is very powerful but strange connection that exists between pairs of particles. Even if they are very far apart, in a way, they are always coordinated. Nature's fundamental building blocks are connected and influence each other instantaneously, as if the space between them does not exist at all. As if two objects can mirror each other without any apparent connection. Einstein called it "spooky action at a distance."

Einstein rejected the idea. He said "*Physics should represent a reality in time and space, free from spooky actions at a distance.*"

Einstein tried to prove it could not be real. You could have situations where the cause and the effect happen at the same time. But if entanglement is not real, cutting-edge technologies could be in jeopardy.

For example, a Quantum computer, and quantum encryption, they depend on entanglement being a fact in the world.

Underlying it all is a profound question: do we live in Einstein's universe of common-sense laws or a bizarre quantum reality that allows spooky connections across space and time?

300 miles off the coast of West Africa, on one of the Canary Islands, a team of physicists set up a remarkable experiment that used almost the entire breadth of the universe to settle the question:

"Is the seemingly impossible phenomenon of quantum entanglement an illusion, or is it actually real?"

Leading the team is Anton Zeilinger. His team used Europe's largest telescopes. Each one simultaneously focused on a different quasar--an extremely distant galaxy.

This light was used to control precise equipment that must be perfectly aligned to make measurements on tiny subatomic particles.

And if that was not tricky enough, the weather on the mountain is often notoriously unpredictable. The team needed perfect conditions for the experiment to work.

With the experiment set up, the team took their positions. David Kaiser worked on the experiment with his colleagues Jason Gallicchio and Andy Friedman for 4 years.

Coordinating it all was Dominik Rauch, who never stopped eating. The experiment is his thesis project and it took years in the making.

Why are physicists so determined to put this bizarre aspect of quantum mechanics to the ultimate test?

To explore the beginning of the story, David Kaiser went to Brussels, the city that Albert Einstein traveled to in 1927 to attend a meeting about a new theory that described the micro-world of atoms and tiny particles -- quantum mechanics.

Quantum mechanics is one of the most amazing intellectual achievements in human history. For the first time, scientists were able to probe a world that was, until then, quite invisible to us. Looking at the world at the scale of atoms, a million times smaller than the width of a human hair. One way to think about scales is that if you take an everyday object, like a soccer ball and you enlarge that soccer ball, so actually you can see the individual atoms, you roughly have to make it the size of the Earth.

And then move into the planet. Then you are in the world of atoms and particles.

It was the nature of fundamental particles, which make up the world we see around us, that Einstein had went to Brussels to discuss. And it was there that Einstein entered into a heated debate that would led to the discovery of quantum entanglement -- a concept that would trouble Einstein for the rest of his life.

So David Kasier went to the place where it all began, the Solvay Institute building in Brussels.

Back in October 1927, where the 5th Solvay Conference was held. An amazing, weeklong series of discussions on really what the world was made of, on the nature of matter and the new quantum theory.

Their experiments were showing that deep inside matter, tiny particles, like atoms and their orbiting electrons, were not solid little spheres.

They seemed fuzzy and undefined. As they dug in, they found things less and less solid. This world was not tiny little bricks that got smaller and smaller.

At some point, the bricks gave way to this mush, and what looked like solidity, solidness, in fact became very confusing and kind of a whole new way of thinking about nature.

The theory of quantum mechanics presented at the Solvay meeting was strange.

They said that a particle like an electron, is not physically real until it is observed, measured by an instrument that can detect it.

Before it is detected, instead of being a solid particle, an electron is just a fuzzy wave, a wave of probability.

These objects, like electrons and atoms, when we describe mathematically their behavior, the only thing we can describe is the probability of being at one place or another.

It is like a wave of all those different possibilities. It is not that the electron is in one place or the other, we just do not know.

It is that the electron really is a combination of every possible place it could be until we look at it.

Quantum mechanics only tells us the probability of a particle's properties, like location.

The laws of nature were no longer definite statements about what is going to happen next.

They were just statements about probabilities. And Einstein felt, that is defeat. We are giving up on the heart of what physics has been, namely, to give a complete description of reality.

For Einstein, the idea that particles only pop into existence when they are observed is akin to magic. It is said he asked,

"*Do you really believe the moon is not there when you are not looking at it?*"

Outside of the formal setting of the conference, he challenged the most vocal supporter of these ideas, the great Danish physicist Niels Bohr.

Einstein would show up to breakfast at the hotel, and Niels Bohr would be there, and Einstein would present his latest challenge.

Niels Bohr would sort of mumble and wonder and confer with his younger colleagues. They would head off to the formal meeting at the institute, and somehow, every night by suppertime, Bohr would have an answer.

One of the observers said that Einstein was like a jack-in-the-box; every day, he would pop up with a new challenge.

And Bohr would flip this way and that, and in the end, by super, have crushed that one, and it would start all over again. To Bohr and his colleagues, quantum mechanics not only explained experimental results, its mathematics were elegant and beautiful.

And since Einstein had not found flaws in their equations, they left the Solva meeting feeling more confident than ever in their ideas.

But Einstein did not give up his conviction that quantum mechanics was flawed. And in his refusal to accept the weird implications of the theory, he would wind up uncovering something even weirder.

In 1933, with the Nazi Party in power in Germany, Einstein chose to settle in America, and he took a position at the Institute for Advanced Study in Princeton, New Jersey.

He recruited two physicists to help him, Nathan Rosen and Boris Podolsky. And in 1935, at afternoon tea, the three men spotted a possible flaw in quantum mechanics that would shake the very foundations of the theory.

They noticed that the mathematics of quantum mechanics led to a seemingly impossible situation. Podolsky would say,

"Well, Professor Einstein, this is very important in your arguments showing that quantum theory is incomplete."

So they got this very animated discussion and what can happen still is, now you have a bunch of scientists discussing, and at some point, someone said, "*Let us write a paper together.*"

So they did. It was called The Einstein Podolsky Rosen Paradox. This paper known today as EPR, argued that the equations of quantum mechanics predicted an impossible connection between particles, a seemingly magical effect.

It would be like having two particles, each hidden under a cup. Looking at one particle mysteriously causes the other to reveal itself, too, with matching properties.

Quantum theory suggested this effect could happen in the real world, for example, with particles of light, photons. The equations implied that a source of photons could create pairs in such a way that when we measure one, causing it to snap out of its fuzzy state, the other mysteriously snaps out of its fuzzy state at the same instant, with correlated properties. The 1935 paper that described this effect has become Einstein's most referenced work of all. It has captivated generations of physicists. Another way to think of the paired particles is to imagine a game of chance that is someone rigged. Suppose you had a pair of quantum dice, you put these two quantum dice in a little cup, then throw them on the table. If you look at them, they show the same number, 6.

If you put them again in the cup, and throw them again, now they both show 3. You put them in again in the cup and throw them again, now they both show 1. The point being, what you see are two random processes, namely, each die showing some number, but these two random processes do the same thing. It is really a mind-boggling thing. How could two particles act in unison, even when they are separated from each other? Essential to EPR argument is that these particles can be separated at an arbitrary distance. One particle could be here on Earth, and the other particle could be in the Andromeda Galaxy, and yet, according to quantum mechanics, a choice to measure something here on Earth is somehow instantaneously affecting what could be said about this other particle.

You cannot go from Earth to Andromeda instantly, and yet that, they argued, is what the equations of quantum mechanics seemed to imply, and that, they said, so much the worse for quantum mechanics. The world simply cannot operate that way. For Einstein, this strange effect conflicted with the most basic concept we use to describe reality, space. For him, objects, particles, everything that exists is located in space. Space, together with time, was the key ingredient in Einstein's theory of special relativity, with his equation

$$E = MC^2$$

Einstein was the master of space-time. He thought that if something happened here, that should not immediately and instantaneously change something that is going on over there, the principle of locality, as we currently call it.

For Einstein it is simply common sense that if objects are separated in space, for one to affect the other, something must travel between them. And that traveling takes time. Quantum particles acting in unison could be explained if they were communicating, one particle instantly sending a signal to the other, telling it what properties it should have. But that would require a signal traveling faster than the speed of light, something Einstein's theory of special relativity had proven impossible.

And it would mean the particles were fuzzy and undefined until the moment they were observed. Instead, Einstein thought the particles should be real all along. They must carry with them a hidden layer of deeper physics that determines their properties from the start. Almost the way that magic tricks, while appearing mysterious, have a hidden explanation. But this hidden physics was missing from quantum theory. So Einstein, Podolsky, and Rosen argued that quantum mechanics was incomplete. Podolsky was very enthusiastic about this project. In fact, he was so enthusiastic that he ran to the "New York Times" and told them the news. So Einstein was really upset with Podolsky, and apparently, he did not speak to him anymore.

When Niels Bohr heard of Einstein's paper, he wrote an obscure response, arguing that one particle could somehow mysteriously influence the other. This seemingly impossible phenomenon became known as quantum entanglement. But Einstein dismissed it as "*spooky actions at a distance.*"

No one could think of an experiment to test whether Einstein or Bohr was correct. But that did not stop physicists and engineers from making use of quantum mechanics to do new things.

In the 1930s and 1940s, the debate around the EPR paper sort of dies down. But, quantum theory actually takes off. The mathematics leads to all kinds of amazing developments.

Entanglement aside, the equations of quantum mechanics enabled the scientists of the Manhattan Project to develop the atomic bomb. And in the years after the Second World War, researchers at Bell Labs in New Jersey used quantum theory to develop one of the first lasers In Bell Labs men experimented with a light once undreamed-of in the natural world. And they built small devices that could control the flow of electricity, transistors, which had a big vital role in the future of mankind, the electronic future.

Transistors became the building blocks of the burgeoning field of electronics. Computers, disc drives, the entire digital revolution soon followed, all made possible by the equations of quantum theory. Einstein's questions about entanglement and what it implied about the incompleteness of quantum mechanics remained unanswered until the 1960s, when a physicist from Northern Ireland made a remarkable breakthrough, John Bell.

Bell was a very talented young physics student, but he quickly grew dissatisfied with he considered almost a kind if dishonesty among his teachers. Bell insisted that Einstein's questions about quantum mechanics had not been addressed.

He got into shouting matches with his professors. He said,

"Do not tell us that Bohr solved all the problems. This really deserves further thought."

Quantum mechanics has been fantastically successful. So it is very intriguing situation that at the, at the foundation of all that impressive success, there are these great doubts. It is a very strange thing that ever since the 1930s, the idea of sitting and thinking hard about the foundations of quantum mechanics has been disreputable among physicists. When people tried to do that, they were kicked out of physics departments.

And so, for someone like Bell, he needed to have a day job doing ordinary particle physics, but at night, hidden away, he could do work on the foundations of quantum mechanics. Bell became a leading particle physicist at CERN, in Geneva. But he continued to explore the debate between Einstein and Bohr. And in 1964, he published an astonishing paper. Bell proved that Bohr's and Einstein's ideas made different predictions. If you could randomly perform one of two possible measurements one each particle, and check how often the results lined up, the answer would reveal whether we lived in Einstein's world, a world that followed common-sense laws, or Bohr's, a world that was deeply strange and allowed spooky quantum connections.

We now know with hindsight this was one of the most significant articles in the history of physics, not just the history of 20th-century physics, but in the history of the field as a whole. But Bell's article appeared in this sort of out-of-the-way journal, in fact, the journal itself folded a few years later. This is not central to the Physics community. It is sort of dutifully sat on library shelves and then forgotten. It literally collected dust on the shelf. A few years later, completely by chance, a brilliant experimental physicist, John Clauser, stumbled upon Bell's article. He thought this is one of the most amazing papers, but where is the experimental evidence?

John word on Bell's theory with fellow physicist Abner Shimony, and at the University of California, Berkeley, started work on experiment to test it. John had a talent for tinkering in the lab and building the parts he needed. He used to rummage around and scavenge and dumpster-drive for old equipment. He knew where to find hidden storage rooms filled with old equipment, which he could raid to salvage spare parts for his experiments. Piece by piece, John Clauser and Stuart Freedman constructed the world's first Bell test experiment. They focused a laser onto calcium atoms, (Picture)

They focused a laser onto calcium atoms, causing them to emit pairs of photons that the equations of quantum theory suggested should be entangled. They recorded whether or not the photons passed through filters on each side and checked how often the answers agreed.

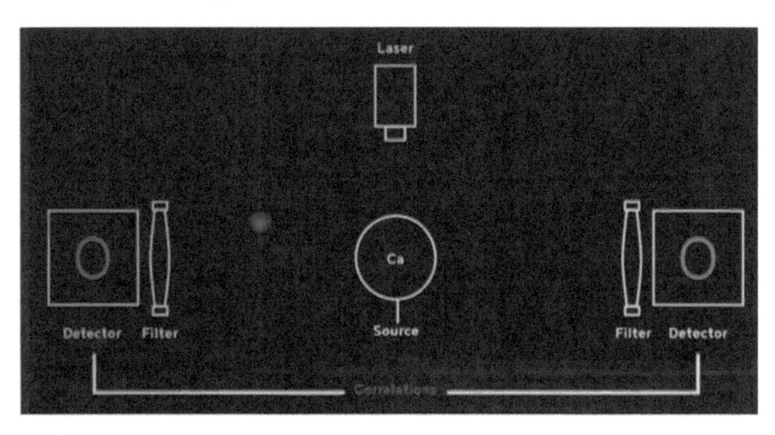

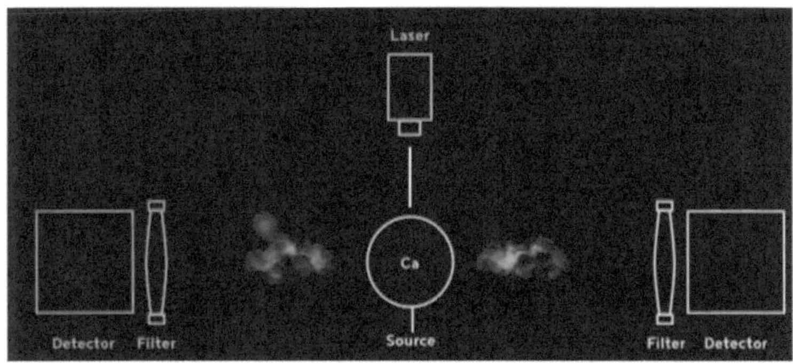

After hundreds of thousands of measurements, if the pairs were more correlated than Einstein's physics predicted, they must be spookily entangled. They saw the stronger correlation characteristic of quantum mechanics, what Bohr's quantum mechanics predicted.

The experiment appeared to show that the spooky connections of quantum entanglement did exist in the natural world. Could it be that the great Albert Einstein was wrong?

Remarkably, the first people to react to this extraordinary result were not the world's leading physicists.

Ronald Reagan's definition of a hippie was someone who dresses like Tarzan, has hair like Jane, and smells like Cheetah.

A small group of free-thinking physicists at the heart of San Francisco's New Age scene got in touch with John. They called themselves the Fundamental Fysiks Group. They spelled physics with an *F*.

Some members would experiment with psychedelic drugs. They were kind of in the flow of the kind of hippie scene. And that group was just mesmerized by the equation of entanglement.

The idea was just to discuss fringe subjects with an open mind. They were trying to link Eastern mysticism with quantum entanglement. They sold a lot of popular textbooks. There were a lot of followers. Their books became bestsellers, like the "*The Tao of Physics*," which highlighted that Eastern philosophy and quantum entanglement both described a deep connectedness of things in the universe.

It was the great cosmic oneness. The group held meetings at the iconic Esalen Institute. It was a marvelous beautiful place where they would discuss all of these ideas. It was right on the Pacific Coast with the overflow from the hot tubs cascading down the cliffs into the Pacific Ocean.

But no useful connections to Eastern mysticism were ever discovered by the group. But it was fun times.

The Fundamental Fysiks Group may not have uncovered the secrets of the "*cosmic oneness*," but in seeing entanglement as central to physics, they were decades ahead of their time.

40 years later, cutting-edge labs around the world are now racing to harness quantum entanglement to create revolutionary new technologies, like quantum computers. In our everyday computers, the fundamental unit of computing is a bit, a binary digit, zero or one. And inside the computer, there is all these transistors, which are turning on and off electric currents.

On is one, off is zero, and these combinations lead to universal computing.

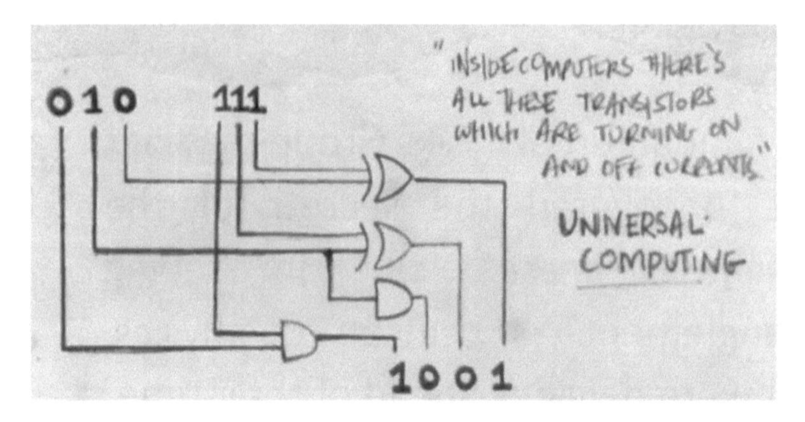

With a quantum computer, you start with a fundamental unit that is not a bit, but a quantum bit, which is not really a zero or a one, but it can be fluid.

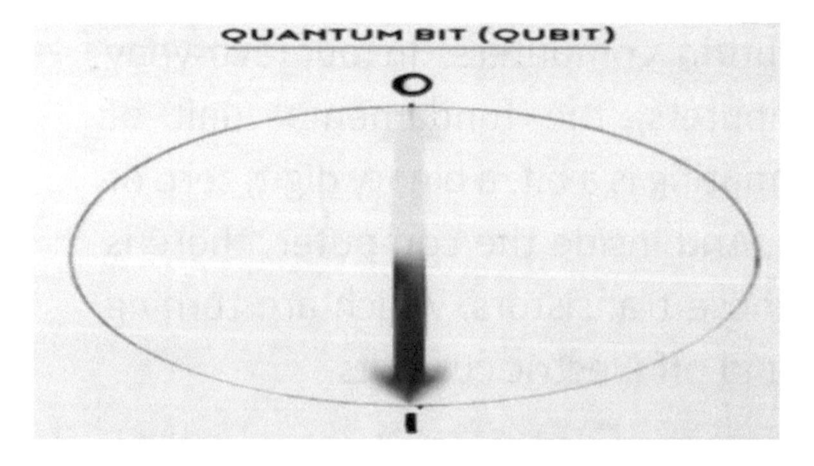

A quantum bit makes use of the fuzziness of the quantum world. A qubit, as it is known, can be zero or one, or a combination of both. A particle or tiny quantum system can be made into a qubit. And today, it is not just pairs of particles that can be entangled, but groups of qubits can be linked with entanglement to create a quantum computer.

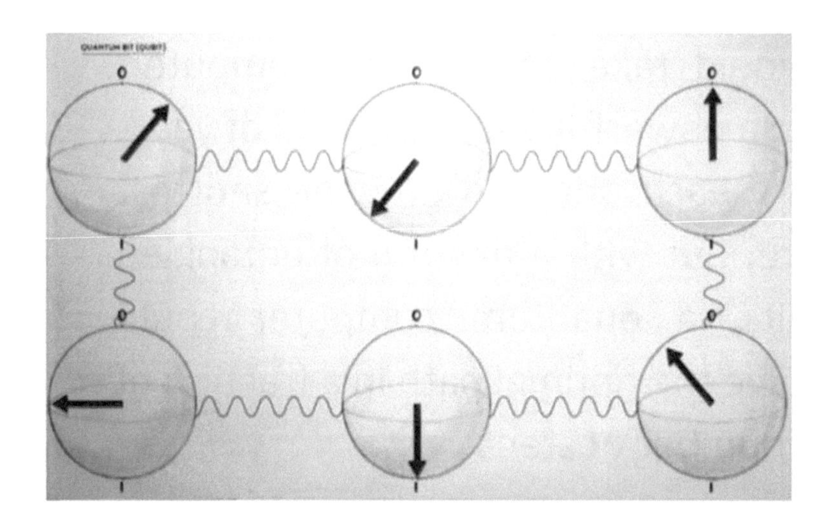

The more qubits, the greater the processing power. By using entangled qubits, quantum computers could tackle real-world problems that traditional computers simply cannot cope with. For example, a salesman has to travel to several cities and wants to find the shortest route.

It sounds easy. But with just 35 cities, there are so many possible routes that it would take an ordinary computer, even a powerful one, hundreds of years to try each one and find the shortest route. But with a handful of entangled qubits, a quantum computer could resolve the optimal path in a fraction of the number of steps.

There is another reason many scientists are racing to create a powerful quantum computer cracking secret codes. In today's modern world, everything from online shopping to covert military communications is protected from hackers using secure digital codes, a process called encryption. But what if hackers could get hold of a quantum computer? A quantum computer could crack our best encryption protocols in minutes, whereas a regular computer, or even a super-computing network today, could not do it in a year's time. But while quantum entanglement may be a threat to traditional encryption, it also offers an even more secure alternative, a communication system that the very laws of physics protect from secret hacking.

But there is a limit to how far quantum signals can be sent through optical fibers. To send signals further, you need a quantum communication satellite. Above the Earth's atmosphere there are fewer obstacles, and quantum particles can travel much further. The satellites can send entangled photons to users on the ground. If an eavesdropper could intercept one of the entangled photons, measure it, and send on a replacement photon, but it would not be an entangled photon. Its properties would not match. It would be clear an eavesdropper was on the line.

In theory, this technique could be used to create a totally secure global communication network.

But there is a problem. What if quantum entanglement "*spooky action at a distance*" is not real after all? It could mean entangled photons are not the path to complete security. The question goes back to Clauser and Freedman's Bell test experiment.

In the years after their pioneering work, physicists began to test possible loopholes in their experiment, ways in which the illusion of entanglement might be created, so the effect might not be so spooky after all. One loophole is especially hard to rule out. In modern Bell test experiments, devices at each side test whether the photons can pass through one of two filters that are randomly chosen, effectively asking one of two questions and checking how often the answers agree.

After thousands of photons, if the results show more agreement than Einstein's physics predicts, the particles must be spookily entangled.

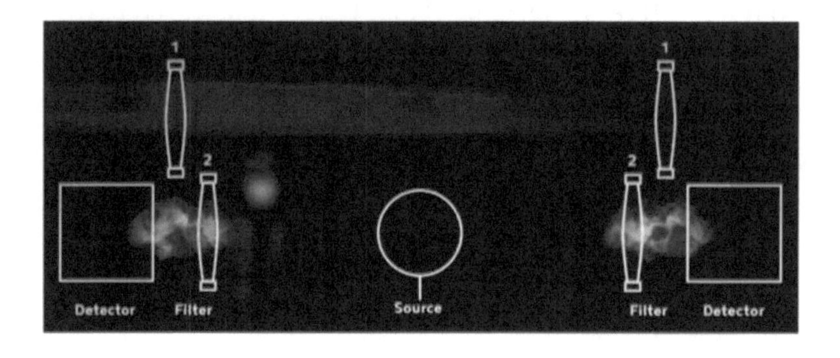

But what if something had mysteriously influenced the equipment so that the choices of the filters were not truly random? Is there any common cause, deep in the past, before you even turn on your device, that could have nudged the questions to be asked and the types of particles to be emitted?

Maybe some strange particle, maybe some force that had not been taken into account, so that what looks like entanglement might indeed be an accident, an illusion. Maybe the world still acts like Einstein thought.

It is this loophole that the team at the high-altitude observatory in the Canary Islands tried to tackle. With quantum mechanics now more established than ever, they put entanglement to the ultimate test, to finally settle the Einstein-Bohr debate beyond all reasonable doubt. The Canary Island team created a giant version of Clauser and Freedman's Bell test, with the entire universe as their lab bench.

In this "*cosmic Bell test*," the source of the entangled particles is about a third of a mile from each of the detectors. The team must send perfectly timed pairs of photons through the air to each side. At the same time, the telescopes will collect light from two extremely far-off, extremely bright galaxies called quasars.

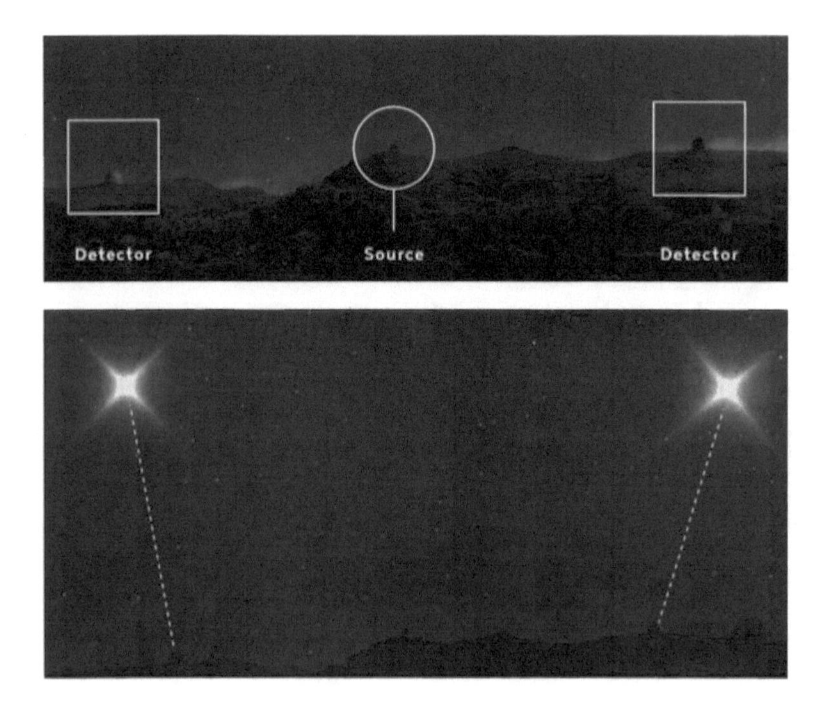

These are among the brightest objects in the sky, emitting light in powerful jets. Random variations in this light will control which filters are used to measure the photon pairs. And since the quasars are so far away the light has been traveling for billions of years to reach Earth, it makes it incredibly unlikely that anything could be influencing the random nature of the test. If the experiment is successful, the team will have tackled the loophole and shown that quantum entanglement is as spooky as Bohr always claimed.

With the telescopes locked on to two different quasars, the team took readings. They did the full cosmic Bell test. It worked. Light from the quasars selected which filters were used to measure the entangled photons.

Two months later, back in Vienna, the team analyzed the experimental data. They saw correlations that corresponded to quantum mechanics. Their results showed entanglement.

And since the light from the quasars controlling the test was nearly eight billion years old, it is extremely unlikely that anything could have affected its random nature.

So this remaining loophole seems to be closed. There is hardly any room left for a kind of alternative, Einstein-like explanation. However, it is still not completely closed. But this experiment have shoved it into such a tiny corner of the cosmos as to make it even more implausible for anything other than entanglement to explain the results.

Accepting that entanglement is a part of the natural world around us has profound implications.

It means we must accept that an action in one place can have an instant effect anywhere in the universe, as if there is no space between them. Or that particles only take on physical properties when we observe them. Or we must accept both.

We are left with conclusions about the universe that make no sense whatsoever. Science is stepping outside of all of our boundaries of common sense. It is almost like being in "*Alice in Wonderland.*" Where everything is very possible.

It was first seen as an unwelcome but unavoidable consequence of quantum mechanics.

Now, after nearly a century of disputes and discoveries, "*spooky action at a distance*" is finally at the heart of modern physics. At the Institute for Advanced Study, where the concept of entanglement was first described, researchers are now using it in their search for a single unified theory of the universe, the holy grail of physics.

Einstein's theories of special and general relativity perfectly describe space, time, and gravity at the largest scales of the universe, while quantum mechanics perfectly describes the tiniest scales. Yet these two theories have never been brought together.

So far, we have not yet had a single complete theory that is both quantum mechanical and reproduces the prediction of Einstein's wonderful theory of general relativity. Maybe the secret is entanglement.

What if space itself is actually created by the tiny quantum world? Just like temperature, warm and cold, consists simply of the movement of atoms inside an object, perhaps space as we know it emerges from networks of entangled quantum particles. It is a mind-blowing idea.

What we are learning these days is that we might have to give up that what Einstein holds sacred, namely, space and time.

Einstein was always thinking,

"*We have little pieces of space and time, and out of this, we build the whole universe*."

In a radical theory, known as the hologrphic universe, space and time are created by entangled quantum particles on a sphere that is infinitely far away. What is happening in space in some sense all described in terms of a screen outside.

The ultimate description of reality resides on this screen. Think of it as kind of quantum bits living on that screen.

And this, like a movie projector, creates an illusion of the three-dimensional reality that we are now experiencing.

It may be impossible to intuitively understand this wild mathematical idea, but it suggests that entanglement could be what forms the true fabric of the universe. The most puzzling element of entanglement that, somehow two points in space can communicate, becomes less of a problem, because space itself has disappeared. In the end, we just have this quantum mechanical world. There is no space anymore. And so in some sense, the paradoxes of entanglement, the EPR paradox disappears into thin air.

To understanding quantum mechanics fully will only happen when we put ourselves on the entanglement side, and we stop privileging the world that we see and start thinking about the world as it actually is.

Science has made enormous progress for centuries by sort of breaking complicated systems down into parts. When we come to a phenomenon like quantum entanglement, that scheme breaks.

When it comes to the bedrock of quantum mechanics, the whole is more than the sum of its parts. The basic motivation is just to learn how nature works.

What is really going on? Einstein said it very nicely. He was not interested in this detailed question or that detailed question. He just wanted to know what Allah's (God) thoughts were when He created the universe.

Made in United States
Troutdale, OR
08/11/2023